Transport Statistics Report

Cars : Make and Model : The Risk of Driver Injury and Car Accident Rates in Great Britain : 1993

February 1995

London: HMSO

15. 1 . 1995

Prepared for publication by STD5 Branch
Directorate of Statistics
Department of Transport

Anthony Craggs
Peter Wilding

GOVERNMENT STATISTICAL SERVICE

A service of statistical information and advice is provided to the government by specialist staff employed in the statistics divisions of individual Departments. Statistics are made generally available through their publications and further information and advice on them can be obtained from the Departments concerned.

Enquiries about the contents of this publication should be made to:

**Directorate of Statistics
Department of Transport
Room B649
Romney House
43 Marsham Street
London SW1P 3PY**

Telephone 071-276-8780

Contents

Page

INTRODUCTION

This publication presents estimates of the risk of driver injury in popular models of car once involved in a two car collision injury accident. It also includes, at a more general level, estimates of accident involvement rates and associated casualty rates analysed, within broad groups, by age and performance of car, and whether the car is privately or company owned. The statistics in this publication are based on personal injury road accident data reported to the Department of Transport by police forces within Great Britain. The publication is in three parts.

The first part relates to the risk of driver injury in popular models of car once involved in two car collision injury accidents, generally known as secondary or passive safety. The estimates are shown in Tables A and B. The second part relates to car accident involvement and casualty rates, generally known as primary or active safety, and estimates are shown in Tables C and D. Part 3 provides summaries of related areas of work currently being developed by the Department.

Structural design is the key factor in secondary safety and is complemented by good design of interior fittings, particularly seat belts, energy absorbing steering wheels and columns, padded dashboards and the provision of airbag supplementary restraints. Together these have great potential for reducing the number of road casualties. By contrast, primary safety - accident avoidance - is mainly influenced by the driver but vehicle design features such as good braking, handling, stability, lighting and driver vision, can all help the driver to avoid accidents.

The first part of the publication summarises the adopted methodology which uses the risk of driver injury as a basis for assessing the secondary safety features of cars, sets out the results in Tables A and B, and summarises the main conclusions. The second part of the publication summarises the accident involvement and casualty rates for particular groups of car, sets out results in Tables C and D, and summarises the main conclusions. The third part outlines progress on work on calculating car safety ratings which only reflect inherent structural safety and the efficiency of safety fittings by excluding the effect of mass in an injury accident. There is also an examination of the aggressivity of cars in injury accidents. This has been determined as the risk of injury to the other driver involved in a collision with a certain model of car. It also includes an analysis of factors affecting the severity of injury to pedestrians hit by cars. This is followed by a section in the form of appendices which set out definitions, and a more detailed description of the source and coverage of the car accident data used in this report for the estimation of car driver injury accident rates.

PART 1

SECONDARY CAR SAFETY (Tables A and B)

Measuring Secondary Car Safety

1.1   One approach for assessing secondary car safety is to simulate typical accidents by crashing a car into a fixed block under controlled conditions and, in some cases, to measure the forces on dummy occupants. Such crash tests are able to control the severity of accident and assess the secondary safety design effects in a simulated collision. The shortcoming of these types of test is that they are representative of only some real life collisions.

1.2   Another approach to measuring secondary safety is to look at real life injury accident data. This is a more complicated approach because, unlike crash tests where severity of accident can be controlled, severity of real-life accidents can be highly variable which makes it difficult to compare injury data for different car models. This is because different car models may be exposed to different types of accident and may be driven predominantly by different groups of drivers.

1.3   The measure of secondary safety for popular models of cars in this publication is based upon the risk of driver injury in two-car injury accidents. This is a relatively straightforward calculation because in every injury accident it is known whether the driver is injured or uninjured because every car must have a driver, except for parked cars which are not included in the analysis. A fuller measure of secondary car safety could also include the risk of injury to front seat passengers. However this is not so straightforward to calculate because only information on injured passengers is recorded by the police, and any risk calculation would have to be based on assumptions about the average number of front seat passengers in models of car to which the number of injured front seat passengers could be compared. Since average front seat occupancy is likely to be variable in different models of car, it was decided to measure secondary safety on the narrower but more reliable definition of driver injury only.

1.4   An assessment of the relative risk of injury in different car models will also be influenced by their rate of involvement in non-injury, ie damage-only, accidents. However, this analysis is necessarily constrained by the availability of only injury accident data.

1.5   In this section, for any injury accident involving a particular model of car, the risk of driver injury is expressed as the proportion of injury accidents in which drivers of that model are injured. These proportions, or rates, are relative rates in the sense that in any two car collision, the relative risk of driver injury not only depends upon the protection offered by the car being driven but also the safety performance of the car being hit. Most popular models of car tend to collide with a similar range of other cars in two car collisions, so for each model the risk of driver injury in that model is being assessed relative to the same average level of safety offered by injury accident involved cars.

1.6   An interesting point to note is the influence of interior, as distinct from exterior, secondary safety features upon the risk of injury in, say, two car collisions. If only one of the cars in a two car collision is fitted with an interior secondary safety device (such as an airbag) then the relative risk of driver injury in that car will be lower. If one of the cars in a two car collision is fitted with better exterior safety devices (such as superior energy absorbing bumpers or crumple zones) the relative risk of injury to both drivers in that particular collision is not affected, since the benefit of these devices is shared equally by both cars. However over the whole range of accidents the beneficial effects upon the risk of driver injury will be relatively concentrated on those models fitted with a superior exterior secondary safety device.

1.7   In general, mass and size of car are closely related. However in larger cars there is usually more room for crumple zones which absorb energy and reduce deceleration forces upon the driver and occupants, reducing their risk of injury. The benefits of size and mass are not available to the same extent in smaller cars. However, mass is a double edged sword to the extent that a heavier car will increase the risk of injury to the driver in the other car involved in the collision.

1.8   In this publication, the risk of driver injury in an injury accident provides the basis for assessing the protection that a car offers its driver. The main methodological problem in comparing driver injury risk in different cars is that the risks are influenced not only by the cars' safety features, but also by the severity of each of the individual accidents in which they are involved. The average severity of accident for each model will be influenced not only by the types of accident they are involved in, but also by the type of driver group which predominates in a particular model of car. For example, some cars may be involved in relatively more injury accidents on motorways at higher speeds than other cars, and some may be driven more by women whose risk of injury once involved in an accident is higher than that for men. A recent study by Dr Jeremy Broughton at the Transport Research Laboratory "The theoretical basis for comparing the accident record of car models" (see References in Appendix 1) confirmed the broad extent to which different models of car are driven by different groups of driver and are involved in accidents on built-up and non built-up roads.

1.9   Only accidents involving two car collisions are considered in the assessment of car secondary safety in order to minimise distortions to the estimates of driver injury risk arising, for example, from a particular model of car having a high proportion of collisions with very large mass vehicles such as Goods Vehicles or Public Service Vehicles. Within the two car accident dataset there will still be some bias in the estimates of driver injury risk for each car model because of the different types of accident and driver involvement for some models of car. These influences are allowed for in a modelling procedure described in Appendix 5. In fact the adjustments made to estimates of driver injury risk to allow for these sources of bias are quite small. Nevertheless the adjusted estimates are the best estimates of secondary car safety derived from the two car collision dataset, and reflect the benefit of the secondary safety protection offered to the driver of a particular car, as well as the benefit of the size or mass of that car.

1.10   In addition to speed a major influence on the severity of accident is the mass of the car. In any two car collision the greater the mass of one of the cars involved in the collision the lower the deceleration force and risk of driver injury in that car. However in that same collision the deceleration force and risk of driver injury in the other lower mass car will usually be higher. Although mass provides safety benefits for the driver of a high mass car it also increases the risk of driver injury in those cars in collision with high mass cars.

1.11   Most models of car collide with a similar size distribution of other cars involved in the collision. Appendix 4 shows the distribution of collisions for each model in relation to the mass group of the other car involved in the collision. At this level of aggregation the distributions are very similar. This leads to the conclusion that in two car accidents a safety rating for each model is calculated as if each car had collided with a car of average mass.

1.12   The rates shown in Table A retain for each particular model the effect of mass as a safety factor. A high mass car, apart from mostly offering more protection to its driver than a small mass car, can also appear to be safer in the statistical rates because it relatively increases the risk of injury in the car that it collides with. Heavy cars are likely to be more aggressive to other cars involved in the collision than light cars, although structural design is also an important factor. The fact that the risk of driver injury is lower in a higher mass car, but that the risk of injury in the other car involved in the collision is higher, presents an interesting paradox. The ratings shown in Table A reflect the benefit of mass to the driver; Part 3 describes a method for assessing a car's safety after allowance has been made for the influence of mass upon the risk of driver injury. This, in principle, removes the benefit of mass from the calculation of the risk of driver injury, so that derivative safety ratings for different models are based on their design, and the effectiveness of their safety fittings, rather than their mass.

1.13   In Table A cars have been classified into four size groups. Within each of the size groups the variation in mass between cars will be less than the variation in mass between cars in different size groups. This means that the risk of driver injury in cars of broadly similar mass within each size group can be compared to assess the protection offered by the structure and design of the car and its secondary safety fittings. But even within the four size groups there remains some mass variation.

Table A : Description

1.14   Table A sets out estimates of the risk of driver injury in two car driver injury accidents, for particular models of car, in four size groups, broadly defined by length of car. The estimates are based on accidents involving two cars during the years 1989, 1990, 1991, 1992 and 1993. Ratings are only published for cars first registered on or after 1st January 1984. The rating for all sizes of car includes those models for which results are not presented individually while the size group averages are based on a weighted average of those models listed in the respective group. The calculation of the risk as the percentage of drivers injured in a specific model of car (say car 1) when involved in an injury accident with any other car (say car 2) can be represented schematically as follows:

$$\frac{X_1 + X_2}{X_1 + X_2 + X_3} \times 100$$

where

X_1 = Number of accidents in which driver injured in car 1 but not in car 2
X_2 = Number of accidents in which driver injured in car 1 and in car 2
X_3 = Number of accidents in which driver injured in car 2 but not in car 1

1.15   Within each size group (Small, Small/Medium, Medium, and Large) estimates are shown for each constituent model of car for all injuries and for fatal or serious injuries. The confidence intervals are also shown for each estimate for each car model. These show the range within which 95 per cent of the time the estimate will lie. For both fatal or serious injuries and all injuries, uncorrected estimates (where no allowances have been made for types of accident or types of driver on the injury rates) are shown in brackets. The corrected estimates are shown alongside their respective confidence intervals. It can be seen that allowing for variation in the type of accidents and type of driver in specific car models does not have a great effect. However some of the allowances are only based on approximate indicators (speed of road as an indicator for speed of accident) and it may be that the available data are not sensitive enough to pick up such effects. The estimates in Table A are based only on two car accidents, which has the benefit of directly eliminating the possibly distorting effects on the ratings, for some models of car, of a higher incidence of single vehicle accidents and collisions with large mass vehicles such as Goods Vehicles and Public Service Vehicles.

1.16   The most significant adjustment was for female drivers. Female drivers are more prone to injury for a given severity of accident and they tend to drive smaller cars than men. The main adjustment therefore tended to improve the uncorrected ratings for small cars where there was a higher incidence of women drivers, but to degrade the uncorrected ratings for large cars where there was a lower incidence of female drivers. Details of the adjustment process for driver influence and type of accident which influence severity of accident can be found in Appendix 5.

Table A : Commentary

1.17   The corrected best estimates of the percentage of drivers injured in any particular model, and their associated confidence intervals, can be used to assess which cars in any group are significantly different from the average risk for the group. Those models with a higher than average risk of driver injury offer a lower than average level of secondary protection, and vice versa for those models with a lower than average risk of driver injury. The estimates have been calculated after allowances have been made for the possible influence of exposure to different types of accident and driver. The specific influences that were looked at were age and sex of driver, point of impact, and speed limit of road as an indicator of severity of accident. More models are identified as being statistically different from their relevant group average for all injuries than for fatal or serious injuries because of the much larger proportions in the samples for all injuries compared to fatal or serious injuries. Consequently the confidence intervals are relatively narrower for all injuries (see Appendix 3).

1.18   Particular car models can be said to be statistically different (95 per cent of the time) from their respective group average if that average lies outside the confidence interval for the particular model. For example, in the grouping of small size cars shown in Table A the Citroen 2CV/Dyane is statistically different from the average risk of driver injury (for all injuries) in all small cars since the confidence interval for the Citroen model (80 to 88 per cent) does not include the estimate of average risk in all small cars (71 per cent). Similarly differences between particular models can be identified if the respective confidence intervals do not overlap. For example, the confidence interval for the Citroen 2CV/Dyane (80 to 88 per cent) does not overlap with the confidence interval for the Renault Clio (58 to 69 per cent).

1.19   The four groups of car shown in Table A have been essentially determined by length of car. Within each of the groups the specific model groupings of car have been determined by considering production line information about specific car models to ensure that the model groupings consist of models with as uniform design as possible. There will be some variation in mass between car models in each of the size groups despite the attempt to standardize the size classification of the four car groups. The estimates of risk of injury for each car model within each size group will reflect variations in mass, and also the secondary safety design and fittings features offered by each model. The important effect of mass and size on risk of injury is clearly shown in Table A. The group average of 71 per cent for small cars falls steadily with group size to 46 per cent for large cars, confirming that, generally, the greater the mass of the car the greater the safety benefit to the driver of that car involved in a collision. In addition, drivers of larger cars are more likely to benefit from more extensive crumple zones. However, it is also true that the greater the mass of the car, the higher the risk of injury in the car that it collides with. This aggressive disbenefit of mass is described further in Part 3 of the publication.

1.20   The four graphs following the commentary on Table A show the safety ratings for groups of car models within each of the four size groups. The safety rating is derived from the risk of injury shown in Table A and relates the risk of injury in a particular model of car to the risk of injury in the overall average car. For example the rating for the Volvo 300 (small/medium cars) is calculated as follows:

All Injury risk (Volvo 300) best estimate   =   55%

All Injury risk (Overall Average car) best estimate   =   63%

$$\text{Safety Rating} = \frac{63\% - 55\%}{63\%} \times 100 = +13\%$$

1.21   This shows that the safety rating for the Volvo 300, in terms of the statistical best estimate, is 13 per cent more than that for the overall average car. A similar calculation is carried out to show how the confidence interval for the safety rating, which includes the best estimate, relates to the overall average car, and these confidence intervals are shown in the four graphs. For example the confidence interval for the Volvo 300 shows that the estimate of the safety rating for this car could be between 10 and 17 per cent better than for the overall average car. The graphs show the confidence interval for each model's rating. Although the ratings have been calculated with reference to the overall average car, the graphs show them in relation to both the overall average and also their respective group average. Within each group of similar sized cars the graphs clearly show groups of cars which, at the 95 per cent level of confidence, are statistically different from their group average, and also from other models within the same group. The following table shows the distribution of these differences within each size group:

Car Size	Below Group Average	Around Group Average	Above Group Average	All Models
Small	3	15	4	22
Small/medium	6	20	10	36
Medium	2	22	6	30
Large	5	9	6	20
All	16	66	26	108

1.22   In each graph, make/models are presented alphabetically within the following three groups; models whose confidence intervals fall entirely below, around and entirely above the average for that size group. The effect of size and mass upon the ratings shows through when the confidence intervals are compared with the overall average. The intervals for small cars mostly fall below the overall average line, but the intervals for large cars all fall above the overall average line.

1.23   The 108 models listed in Table A include 20 pairs where the model name has been retained by the manufacturer for the newer replacement model, but where the newer versions of that name include different design features and are therefore treated separately in the analysis. The replacement models generally provide at least as much protection to their drivers as their predecessors, and the latest versions of the following five models provide significantly better protection: Ford Escort, Rover 200/400, Vauxhall Astra, Volkswagen Golf and Toyota Carina. This pattern extends to the new versions of other models, and also to new models, for which ratings are also generally better than the average for their respective size groups.

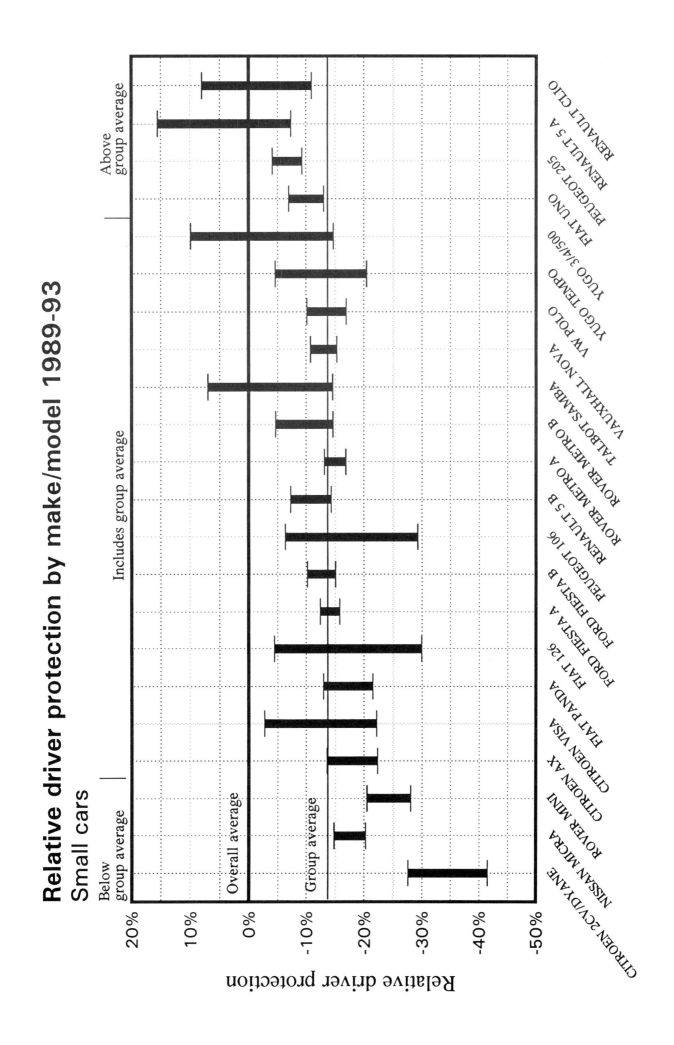

Relative driver protection by make/model 1989-93
Small cars

7

Relative driver protection by make/model 1989-93

Small/medium cars

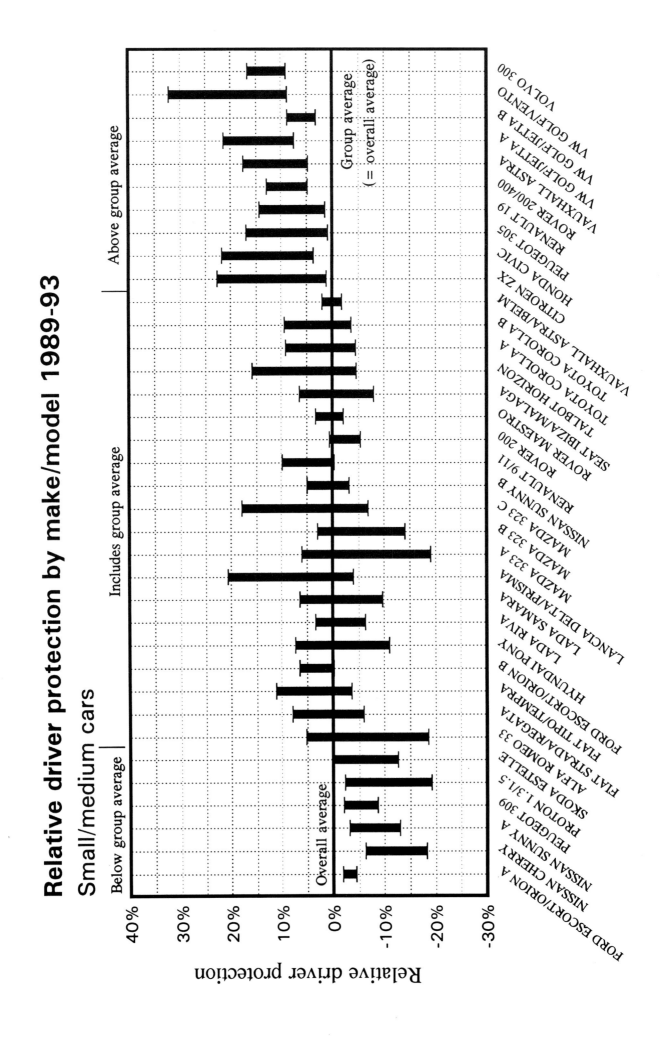

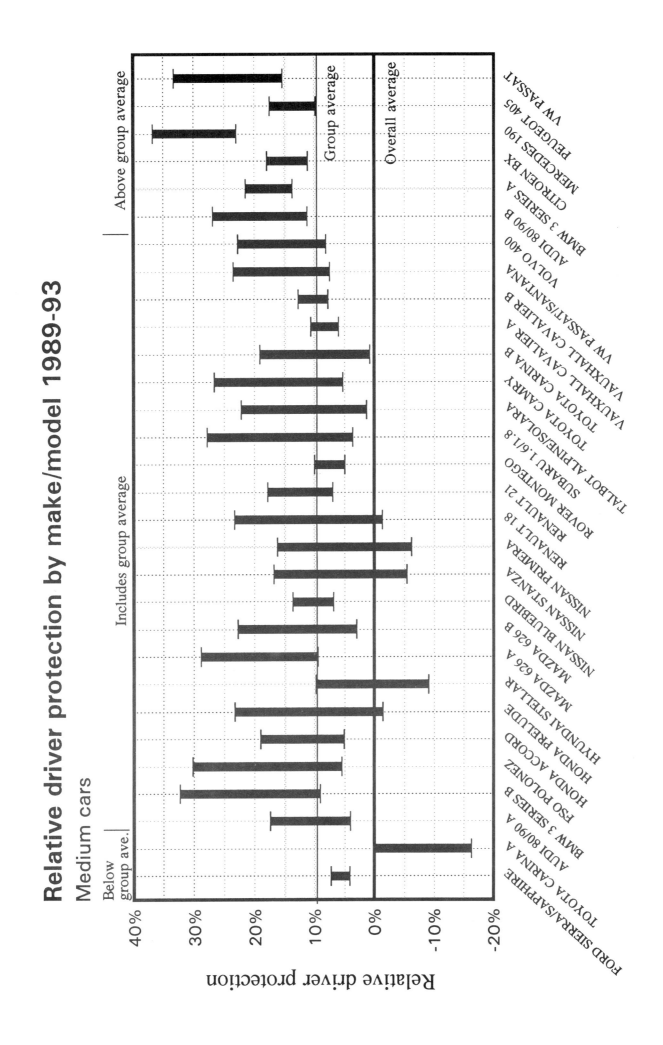

Relative driver protection by make/model 1989-93

Medium cars

Relative driver protection by make/model 1989-93

Large cars

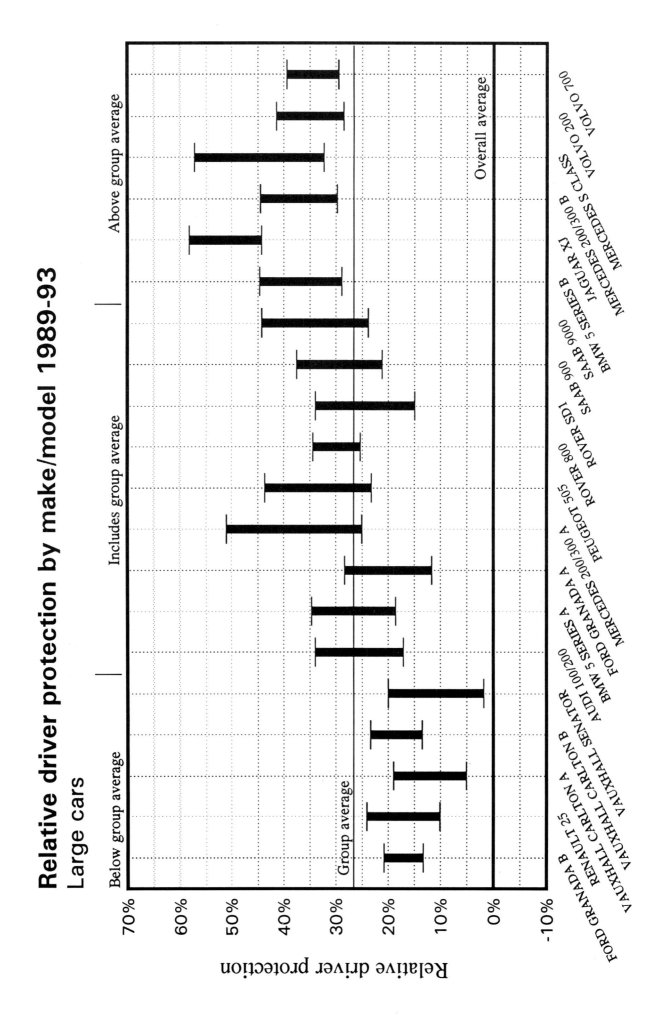

GUIDE TO MAKE/MODELS IN GRAPHS

Car size/model	Registration dates	Car size/model	Registration dates
SMALL		**SMALL/MEDIUM**	
CITROEN 2CV/DYANE	Jan 84 - Jul 90	ALFA ROMEO 33	Jan 84 - Dec 93
CITROEN AX	Jun 87 - Dec 93	CITROEN ZX	Jun 91 - Dec 93
CITROEN VISA	Jan 84 - Jul 88	FIAT STRADA/REGATA	Jan 84 - Jun 88
FIAT 126	Jan 84 - Dec 92	FIAT TIPO/TEMPRA	Jul 88 - Dec 93
FIAT PANDA	Jan 84 - Dec 93	FORD ESCORT/ORION A	Jan 84 - Aug 90
FIAT UNO	Jan 84 - Dec 93	FORD ESCORT/ORION B	Sep 90 - Dec 93
FORD FIESTA A	Jan 84 - Mar 89	HONDA CIVIC	Oct 87 - Oct 91
FORD FIESTA B	Apr 89 - Dec 93	HYUNDAI PONY	Oct 85 - Aug 90
NISSAN MICRA	Jan 84 - Dec 92	LADA RIVA	Jan 84 - Dec 93
PEUGEOT 106	Oct 91 - Dec 93	LADA SAMARA	Nov 87 - Dec 93
PEUGEOT 205	Jan 84 - Dec 93	LANCIA DELTA/PRISMA	Jan 84 - Dec 93
RENAULT 5 A	Jan 84 - Jan 85	MAZDA 323 A	Jan 84 - Aug 85
RENAULT 5 B	Feb 85 - Dec 93	MAZDA 323 B	Sep 85 - Sep 89
RENAULT CLIO	Mar 91 - Dec 93	MAZDA 323 C	Oct 89 - Dec 93
ROVER METRO A	Jan 84 - Mar 90	NISSAN CHERRY	Jan 84 - Aug 86
ROVER METRO B	Apr 90 - Dec 93	NISSAN SUNNY A	Jan 84 - Aug 86
ROVER MINI	Jan 84 - Dec 93	NISSAN SUNNY B	Sep 86 - Jan 91
TALBOT SAMBA	Jan 84 - Sep 86	PEUGEOT 305	Jan 84 - Jul 88
VAUXHALL NOVA	Jan 84 - Mar 93	PEUGEOT 309	Feb 86 - Mar 93
VOLKSWAGEN POLO	Jan 84 - Dec 93	PROTON 1.3/1.5	Mar 89 - Dec 93
YUGO 3/4/500	Jan 84 - Dec 91	RENAULT 19	Feb 89 - Dec 93
YUGO TEMPO	Jan 84 - Dec 93	RENAULT 9/11	Jan 84 - Jan 89
		ROVER 200	Jun 84 - Sep 89
		ROVER 200/400	Oct 89 - Dec 93
		ROVER MAESTRO	Jan 84 - Dec 93
		SEAT IBIZA/MALAGA	Oct 85 - Sep 93
		SKODA ESTELLE	Jan 84 - Jul 90
MEDIUM		TALBOT HORIZON	Jan 84 - Dec 85
AUDI 80/90 A	Jan 84 - Oct 86	TOYOTA COROLLA A	Jan 84 - Aug 87
AUDI 80/90 B	Nov 86 - Dec 93	TOYOTA COROLLA B	Sep 87 - Jul 92
BMW 3 SERIES A	Jan 84 - Mar 91	VAUXHALL ASTRA/BELMONT	Oct 84 - Sep 91
BMW 3 SERIES B	Apr 91 - Dec 93	VAUXHALL ASTRA	Oct 91 - Dec 93
CITROEN BX	Jan 84 - Dec 93	VOLKSWAGEN GOLF/JETTA A	Jan 84 - Feb 84
FORD SIERRA/SAPPHIRE	Jan 84 - Dec 93	VOLKSWAGEN GOLF/JETTA B	Mar 84 - Jan 92
FSO POLONEZ	Jan 84 - Dec 91	VOLKSWAGEN GOLF/VENTO	Feb 92 - Dec 93
HONDA ACCORD	Oct 85 - Sep 91	VOLVO 300	Jan 84 - Dec 91
HONDA PRELUDE	Jan 84 - Mar 92		
HYUNDAI STELLAR	Jun 84 - Dec 91		
MAZDA 626 A	Jan 84 - Sep 87	**LARGE**	
MAZDA 626 B	Oct 87 - Jan 92	AUDI 100/200	Jan 84 - Dec 93
MERCEDES 190	Jan 84 - Sep 93	BMW 5 SERIES A	Jan 84 - May 88
NISSAN BLUEBIRD	Mar 86 - Aug 90	BMW 5 SERIES B	Jun 88 - Dec 93
NISSAN PRIMERA	Sep 90 - Dec 93	FORD GRANADA A	Jan 84 - Apr 85
NISSAN STANZA	Jan 84 - Dec 86	FORD GRANADA B	May 85 - Dec 93
PEUGEOT 405	Jan 88 - Dec 93	JAGUAR XJ	Oct 86 - Dec 93
RENAULT 18	Jan 84 - May 86	MERCEDES S CLASS	Jan 84 - Sep 91
RENAULT 21	Jun 86 - Dec 93	MERCEDES 200/300 A	Jan 84 - Sep 85
ROVER MONTEGO	Apr 84 - Dec 93	MERCEDES 200/300 B	Oct 85 - Dec 93
SUBARU 1.6/1.8	Nov 84 - Dec 91	PEUGEOT 505	Jan 84 - Dec 91
TALBOT ALPINE/SOLARA	Jan 84 - Dec 86	RENAULT 25	Jul 84 - Jan 93
TOYOTA CAMRY	Jan 84 - Dec 86	ROVER 800	Jul 86 - Dec 93
TOYOTA CARINA A	Apr 84 - Feb 88	ROVER SD1	Jan 84 - Jun 86
TOYOTA CARINA B	Mar 88 - Apr 92	SAAB 900	Jan 84 - Sep 93
VAUXHALL CAVALIER A	Jan 84 - Sep 88	SAAB 9000	Oct 85 - Dec 93
VAUXHALL CAVALIER B	Oct 88 - Dec 93	VAUXHALL CARLTON A	Jan 84 - Oct 86
VOLKSWAGEN PASSAT/SANTANA	Jan 84 - May 88	VAUXHALL CARLTON B	Nov 86 - Dec 93
VOLKSWAGEN PASSAT	Jun 88 - Dec 93	VAUXHALL SENATOR	Sep 87 - Dec 93
VOLVO 400	Jun 87 - Dec 93	VOLVO 200	Jan 84 - Dec 93
		VOLVO 700	Jan 84 - Jul 91

Table A Risk of injury to car drivers involved in two car injury accidents: by size and make/model of car: 1989 to 1993

		Percentage of drivers injured when involved in an injury accident[1]					
		Injury severity					
		Fatal or serious			All		
Car size/model[2]	Registration dates	Corrected[3]	95% C.I.[4]	Uncorrected	Corrected[3]	95% C.I.[4]	Uncorrected
SMALL							
CITROEN 2CV/DYANE	Jan 84 - Jul 90[5]	10	[7 , 14]	(11)	85	[80 , 88]	(87)
CITROEN AX	Jun 87 - Dec 93	8	[7 , 10]	(8)	74	[71 , 77]	(75)
CITROEN VISA	Jan 84 - Jul 88[5]	12	[8 , 16]	(13)	71	[64 , 76]	(73)
FIAT 126	Jan 84 - Dec 92	12	[8 , 19]	(12)	74	[65 , 81]	(81)
FIAT PANDA	Jan 84 - Dec 93	9	[7 , 10]	(8)	73	[71 , 76]	(76)
FIAT UNO	Jan 84 - Dec 93	8	[7 , 9]	(8)	69	[67 , 71]	(69)
FORD FIESTA	Jan 84 - Mar 89	9	[8 , 9]	(8)	71	[70 , 72]	(73)
FORD FIESTA	Apr 89 - Dec 93	7	[6 , 8]	(7)	70	[69 , 72]	(72)
NISSAN MICRA	Jan 84 - Dec 92	8	[7 , 10]	(8)	74	[72 , 75]	(77)
PEUGEOT 106	Oct 91 - Dec 93	7	[4 , 12]	(7)	74	[67 , 81]	(77)
PEUGEOT 205	Jan 84 - Dec 93	8	[7 , 9]	(9)	67	[65 , 68]	(68)
RENAULT 5	Jan 84 - Jan 85	9	[6 , 13]	(9)	60	[53 , 67]	(64)
RENAULT 5	Feb 85 - Dec 93	8	[7 , 10]	(8)	69	[67 , 72]	(71)
RENAULT CLIO	Mar 91 - Dec 93	8	[5 , 11]	(7)	64	[58 , 69]	(68)
ROVER METRO	Jan 84 - Mar 90	9	[8 , 9]	(8)	72	[71 , 73]	(75)
ROVER METRO	Apr 90 - Dec 93	8	[6 , 10]	(8)	69	[66 , 72]	(71)
ROVER MINI	Jan 84 - Dec 93	11	[9 , 13]	(10)	78	[75 , 80]	(82)
TALBOT SAMBA	Jan 84 - Sep 86[5]	7	[4 , 11]	(7)	65	[58 , 72]	(70)
VAUXHALL NOVA	Jan 84 - Mar 93	8	[7 , 9]	(8)	71	[69 , 72]	(74)
VOLKSWAGEN POLO	Jan 84 - Dec 93	8	[7 , 9]	(8)	71	[69 , 73]	(74)
YUGO 3/4/500	Jan 84 - Dec 91	13	[9 , 19]	(12)	64	[56 , 72]	(64)
YUGO TEMPO	Jan 84 - Dec 93	5	[3 , 8]	(4)	71	[65 , 75]	(72)
ALL SMALL		8		(8)	71		(73)
SMALL/MEDIUM							
ALFA ROMEO 33	Jan 84 - Dec 93	8	[5 , 13]	(10)	67	[59 , 74]	(63)
CITROEN ZX	Jun 91 - Dec 93	7	[4 , 11]	(7)	55	[48 , 62]	(57)
FIAT STRADA/REGATA	Jan 84 - Jun 88[5]	8	[6 , 11]	(8)	62	[58 , 66]	(58)
FIAT TIPO/TEMPRA	Jul 88 - Dec 93	5	[4 , 8]	(6)	60	[56 , 65]	(57)
FORD ESCORT/ORION	Jan 84 - Aug 90	7	[7 , 8]	(7)	65	[64 , 65]	(61)
FORD ESCORT/ORION	Sep 90 - Dec 93	5	[4 , 6]	(5)	61	[58 , 63]	(57)
HONDA CIVIC	Oct 87 - Oct 91	6	[4 , 9]	(7)	55	[49 , 60]	(60)
HYUNDAI PONY	Oct 85 - Aug 90	9	[7 , 13]	(10)	64	[58 , 69]	(64)
LADA RIVA[6]	Jan 84 - Dec 93	7	[6 , 9]	(8)	64	[60 , 67]	(61)
LADA SAMARA	Nov 87 - Dec 93	5	[4 , 8]	(6)	64	[59 , 69]	(62)
LANCIA DELTA/PRISMA	Jan 84 - Dec 93	9	[6 , 15]	(9)	58	[50 , 65]	(53)
MAZDA 323	Jan 84 - Aug 85	9	[5 , 14]	(9)	67	[59 , 75]	(68)
MAZDA 323	Sep 85 - Sep 89	7	[5 , 10]	(8)	66	[61 , 71]	(67)
MAZDA 323	Oct 89 - Dec 93	5	[3 , 9]	(6)	59	[51 , 67]	(62)
NISSAN CHERRY	Jan 84 - Aug 86[5]	10	[8 , 13]	(9)	70	[66 , 74]	(70)
NISSAN SUNNY	Jan 84 - Aug 86	9	[7 , 11]	(8)	68	[65 , 71]	(65)
NISSAN SUNNY	Sep 86 - Jan 91	6	[5 , 8]	(6)	62	[59 , 65]	(61)
PEUGEOT 305	Jan 84 - Jul 88[5]	6	[4 , 9]	(7)	57	[52 , 62]	(53)
PEUGEOT 309	Feb 86 - Mar 93	8	[7 , 9]	(8)	66	[64 , 68]	(64)
PROTON 1.3/1.5	Mar 89 - Dec 93	7	[5 , 11]	(8)	70	[64 , 75]	(68)
RENAULT 19	Feb 89 - Dec 93	6	[5 , 8]	(7)	58	[54 , 62]	(56)
RENAULT 9/11	Jan 84 - Jan 89[5]	8	[7 , 10]	(8)	60	[56 , 63]	(60)
ROVER 200	Jun 84 - Sep 89	7	[6 , 8]	(8)	64	[62 , 66]	(62)
ROVER 200/400	Oct 89 - Dec 93	7	[6 , 8]	(8)	57	[54 , 60]	(55)
ROVER MAESTRO	Jan 84 - Dec 93	6	[5 , 7]	(6)	62	[60 , 64]	(60)
SEAT IBIZA/MALAGA	Oct 85 - Sep 93	6	[4 , 9]	(6)	63	[59 , 68]	(63)
SKODA ESTELLE	Jan 84 - Jul 90[5]	8	[6 , 11]	(9)	67	[63 , 70]	(68)
TALBOT HORIZON	Jan 84 - Dec 85[5]	5	[3 , 9]	(6)	59	[53 , 65]	(61)
TOYOTA COROLLA	Jan 84 - Aug 87	8	[6 , 11]	(8)	61	[57 , 65]	(60)
TOYOTA COROLLA	Sep 87 - Jul 92	7	[5 , 9]	(7)	61	[57 , 65]	(62)
VAUXHALL ASTRA/BELMONT	Oct 84 - Sep 91	7	[6 , 7]	(7)	63	[61 , 64]	(60)
VAUXHALL ASTRA	Oct 91 - Dec 93	3	[2 , 5]	(3)	56	[52 , 60]	(54)
VOLKSWAGEN GOLF/JETTA	Jan 84 - Feb 84	5	[3 , 7]	(5)	54	[49 , 58]	(53)
VOLKSWAGEN GOLF/JETTA	Mar 84 - Jan 92	6	[5 , 7]	(6)	59	[57 , 61]	(58)
VOLKSWAGEN GOLF/VENTO	Feb 92 - Dec 93	3	[2 , 7]	(3)	50	[42 , 57]	(48)
VOLVO 300	Jan 84 - Dec 91[5]	6	[5 , 7]	(7)	55	[52 , 57]	(56)
ALL SMALL/MEDIUM		7		(7)	63		(60)

1 Excluding accidents in which neither driver was injured.
2 Models are listed under the current market name of the manufacturer.
3 Corrected for selected accident circumstances such as road type and driver age. Uncorrected rates are shown in brackets.
4 The probability is 95% that the true rate lies within the range.
5 Approximate withdrawal date. All cars registered up to December 1993 are included.
6 Includes earlier 1200, 1300, 1500 and 1600 models.

Table A (cont'd)   Risk of injury to car drivers involved in two car injury accidents: by size and make/model of car: 1989 to 1993

		Percentage of drivers injured when involved in an injury accident[1]					
		Injury severity					
		Fatal or serious			All		
Car size/model[2]	Registration dates	Corrected[3]	95% C.I.[4]	Uncorrected	Corrected[3]	95% C.I.[4]	Uncorrected
MEDIUM							
AUDI 80/90	Jan 84 - Oct 86	7	[5 , 9]	(7)	56	[52 , 60]	(53)
AUDI 80/90	Nov 86 - Dec 93	4	[2 , 6]	(5)	51	[46 , 56]	(49)
BMW 3 SERIES	Jan 84 - Mar 91	6	[5 , 8]	(6)	52	[49 , 54]	(49)
BMW 3 SERIES	Apr 91 - Dec 93	4	[2 , 7]	(4)	50	[42 , 57]	(46)
CITROEN BX	Jan 84 - Dec 93	5	[4 , 6]	(6)	53	[51 , 56]	(50)
FORD SIERRA/SAPPHIRE	Jan 84 - Dec 93	6	[6 , 7]	(6)	59	[58 , 60]	(53)
FSO POLONEZ	Jan 84 - Dec 91[5]	6	[3 , 10]	(6)	51	[44 , 59]	(51)
HONDA ACCORD	Oct 85 - Sep 91	8	[6 , 10]	(8)	55	[51 , 59]	(54)
HONDA PRELUDE	Jan 84 - Mar 92	5	[3 , 10]	(5)	56	[48 , 63]	(56)
HYUNDAI STELLAR	Jun 84 - Dec 91	5	[3 , 9]	(5)	63	[56 , 68]	(57)
MAZDA 626	Jan 84 - Sep 87	5	[3 , 8]	(5)	51	[44 , 57]	(49)
MAZDA 626	Oct 87 - Jan 92	8	[5 , 12]	(9)	55	[48 , 61]	(54)
MERCEDES 190	Jan 84 - Sep 93	4	[3 , 6]	(4)	44	[40 , 48]	(42)
NISSAN BLUEBIRD	Mar 86 - Aug 90[5]	6	[5 , 7]	(5)	56	[54 , 58]	(51)
NISSAN PRIMERA	Sep 90 - Dec 93	5	[3 , 9]	(5)	60	[52 , 66]	(57)
NISSAN STANZA	Jan 84 - Dec 86[5]	9	[6 , 14]	(7)	59	[52 , 66]	(59)
PEUGEOT 405	Jan 88 - Dec 93	5	[4 , 6]	(6)	54	[52 , 56]	(51)
RENAULT 18	Jan 84 - May 86[5]	10	[7 , 16]	(11)	56	[48 , 63]	(55)
RENAULT 21	Jun 86 - Dec 93	6	[5 , 8]	(7)	55	[51 , 58]	(52)
ROVER MONTEGO	Apr 84 - Dec 93	6	[6 , 7]	(6)	58	[56 , 59]	(53)
SUBARU 1.6/1.8	Nov 84 - Dec 91	5	[3 , 9]	(7)	53	[45 , 60]	(55)
TALBOT ALPINE/SOLARA	Jan 84 - Dec 86[5]	6	[4 , 10]	(6)	55	[49 , 62]	(54)
TOYOTA CAMRY	Jan 84 - Dec 86	6	[4 , 10]	(7)	53	[46 , 59]	(48)
TOYOTA CARINA	Apr 84 - Feb 88	6	[4 , 9]	(6)	68	[62 , 73]	(64)
TOYOTA CARINA	Mar 88 - Apr 92	6	[4 , 9]	(7)	56	[51 , 62]	(53)
VAUXHALL CAVALIER	Jan 84 - Sep 88	6	[6 , 7]	(6)	57	[56 , 59]	(52)
VAUXHALL CAVALIER	Oct 88 - Dec 93	5	[5 , 6]	(5)	56	[55 , 58]	(51)
VOLKSWAGEN PASSAT/SANTANA	Jan 84 - May 88	6	[4 , 9]	(7)	53	[48 , 58]	(52)
VOLKSWAGEN PASSAT	Jun 88 - Dec 93	4	[2 , 6]	(4)	47	[42 , 53]	(43)
VOLVO 400	Jun 87 - Dec 93	6	[4 , 8]	(7)	53	[48 , 57]	(54)
ALL MEDIUM		6		(6)	57		(52)
LARGE							
AUDI 100/200	Jan 84 - Dec 93	3	[2 , 5]	(4)	47	[41 , 52]	(43)
BMW 5 SERIES	Jan 84 - May 88	4	[3 , 7]	(4)	46	[41 , 51]	(43)
BMW 5 SERIES	Jun 88 - Dec 93	4	[3 , 7]	(4)	39	[35 , 45]	(35)
FORD GRANADA	Jan 84 - Apr 85	6	[4 , 8]	(5)	50	[45 , 55]	(44)
FORD GRANADA	May 85 - Dec 93	6	[5 , 7]	(6)	52	[50 , 54]	(48)
JAGUAR XJ	Oct 86 - Dec 93	3	[2 , 5]	(3)	30	[26 , 35]	(28)
MERCEDES S CLASS	Jan 84 - Sep 91	4	[2 , 6]	(4)	34	[27 , 42]	(30)
MERCEDES 200/300	Jan 84 - Sep 85	2	[1 , 6]	(2)	38	[31 , 47]	(35)
MERCEDES 200/300	Oct 85 - Dec 93	4	[3 , 6]	(4)	39	[35 , 44]	(38)
PEUGEOT 505	Jan 84 - Dec 91[5]	3	[1 , 5]	(3)	42	[35 , 48]	(41)
RENAULT 25	Jul 84 - Jan 93	5	[4 , 7]	(6)	52	[48 , 56]	(46)
ROVER 800	Jul 86 - Dec 93	4	[3 , 5]	(4)	44	[41 , 47]	(38)
ROVER SD1	Jan 84 - Jun 86[5]	5	[3 , 8]	(5)	47	[41 , 53]	(44)
SAAB 900	Jan 84 - Sep 93	3	[2 , 5]	(4)	44	[39 , 49]	(41)
SAAB 9000	Oct 85 - Dec 93	4	[2 , 6]	(4)	41	[35 , 48]	(39)
VAUXHALL CARLTON	Jan 84 - Oct 86	5	[4 , 8]	(5)	55	[51 , 59]	(51)
VAUXHALL CARLTON	Nov 86 - Dec 93	5	[4 , 6]	(5)	51	[48 , 54]	(47)
VAUXHALL SENATOR	Sep 87 - Dec 93	5	[3 , 8]	(5)	56	[50 , 61]	(50)
VOLVO 200	Jan 84 - Dec 93	4	[3 , 5]	(4)	41	[37 , 45]	(41)
VOLVO 700	Jan 84 - Jul 91[5]	3	[3 , 5]	(4)	41	[38 , 44]	(39)
ALL LARGE		4		(5)	46		(42)
ALL SIZES		7		(7)	63		(60)

1 Excluding accidents in which neither driver was injured.
2 Models are listed under the current market name of the manufacturer.
3 Corrected for selected accident circumstances such as road type and driver age. Uncorrected rates are shown in brackets.
4 The probability is 95% that the true rate lies within the range.
5 Approximate withdrawal date. All cars registered up to December 1993 are included.

Table B : Description

1.24   The effects of various accident characteristics on the risk of driver injury when involved in a two car injury accident are shown in Table B. The percentages are derived using a statistical modelling technique. The most important consequence of analysing the data in this way is that the estimates of injury risk are independent of each other. For example, the estimates of injury risk on different road types are the best available estimates of the specific effect of road type, after the influences of variations in the other effects listed have been removed. The risks shown in the table relate to the average specific effects of those factors in injury accidents. Further explanation and information on the analyses are given in Appendix 5.

1.25   Table A shows the percentage of car drivers injured when involved in an injury accident for individual models of car. The percentages are derived from statistical models similar to those used for Table B but extended to allow for the effect of model of car.

Table B : Commentary

1.26   The rates shown relate to the average specific effects of those factors in injury accidents, after correcting for the influences of variations in the other factors listed. For example, the risks of injury tabulated for different sizes of car are corrected to allow for differences in the age and sex of their drivers.

1.27   Of the factors considered, those having the most influence on the risk of death or serious injury when involved in an injury accident are the speed limit of the road, the first point of impact and driver age. The most severe accidents occur on 60 mph roads, where the risk of death or serious injury to the driver is 3 times higher than on a 20 or 30 mph road. The risk of death or serious injury is about one quarter lower on a 70 mph road than on a 60 mph road, which is perhaps a reflection of the improved segregation of traffic on multi-carriageway roads. A driver is about three times more likely to be killed or seriously injured in a frontal or side impact collision than when their car is hit from behind. However a rear impact is more likely to result in slight injuries; these are likely to be whiplash neck injuries associated with either poorly adjusted or non-existent head restraints.

1.28   Age is not a large influence on the overall risk of injury when involved in an accident, although there is increased susceptibility to death or serious injury in the over 55 age group. Women are less likely to be killed when involved in an accident, which may indicate involvement at lower impact speeds. However they are about 40 per cent more likely to be injured when involved in a two car accident than men. This could be due to the different size of women, and the fact that women are more likely to sit closer to the steering wheel making them more prone to injury.

Table B Risk of injury to drivers involved in two car injury accidents: by various factors: 1989 to 1993

| | Percentage of drivers injured when involved in a two car injury accident[1] | | |
| | Injury severity[2] | | |
Factor	Fatal	Fatal or Serious	All
Speed limit of road (mph)			
20 or 30	0.2	5	59
40 or 50	0.7	8	61
60	2.0	15	71
70	2.0	11	60
Sex of driver			
Male	0.5	7	54
Female	0.3	8	75
Age of driver			
17 - 24	0.4	7	63
25 - 34	0.3	7	61
35 - 54	0.4	7	62
55 or more	1.0	10	68
Size of car			
Small	0.6	9	72
Small/medium	0.4	7	63
Medium	0.3	6	56
Large	0.2	5	45
First point of impact			
Front	0.5	8	53
Back	0.1	3	83
Offside	0.8	9	70
Nearside	0.9	8	64
All accidents	0.4	7	63

1 Excluding accidents in which neither driver was injured.
2 Corrected for influences of variations in the other factors listed.

Summary

1.29   The risks of driver injury in injury accidents for particular popular models of car are presented as rates or proportions of driver injuries in Table A. The rates reflect the secondary protection offered to the driver, but also reflect the influence of mass in collisions. A problem in comparing rates, to assess the relative secondary safety of cars, is the influence of variability in accident severity. In each collision the relative severity of accident for each vehicle depends on the differences in mass between the vehicles, and also the type of accident and drivers involved. The variation in risk of driver injury due to the latter influences is accounted for by a modelling procedure which suggests that only minor adjustments should be made to the risks of driver injury in each model of car. The assumption that each model of car has a similar distribution of collisions in terms of the size of the other car involved implies that each car is effectively being assessed against a car of average mass. Any variation in driver injury risk due to differences in the size distribution of two car collisions is therefore likely to be minimal. The ratings in Table A retain the influence of mass, and support the indication that heavier cars tend to be safer cars for their occupants. However, within size groups where car mass is less variable, there are discernible differences in risk of driver injury. It is also clear that heavier cars tend to increase the risk of injury in the cars they collide with and this is described further in Part 3.

1.30   The estimation of injury risk, and derivative car safety ratings, from road accident injury data is inevitably an evolutionary process. Methods of analysis are not static, and the amount of accident data available for a particular model of car gradually accumulates while a model is still active on the roads. Nevertheless the data records used to date do highlight some vehicles which are statistically better or worse than others, and provide some indication of relative safety. The data are recognised as a useful tool for informing car consumers about the safety features of cars, and for promoting the importance of car safety as a means of saving lives and avoiding serious injury.

PART 2

PRIMARY CAR SAFETY (Tables C and D)

Measuring Primary Car Safety

2.1   The risk of being involved in accidents is influenced mainly by driver behaviour, because the vast majority of accidents are initiated by driver error. However, vehicle characteristics such as braking, handling, lighting and drivers' fields of vision, known as primary safety factors, can influence the effects of driver error on accident risk, as can the road and traffic management infrastructure. In addition to these factors, the number of road injury accidents involving a particular type of vehicle will depend on the number of such vehicles on the road and the average mileage driven. These influences are known collectively as measures of exposure. Part 2 examines the involvement and casualty rates of different groups of car in injury accidents but the data do not fully account for exposure.

2.2   The Department obtains information on vehicle populations through analysis of Driver and Vehicle Licensing Agency (DVLA) registrations data. The stock figures used in these analyses are end of year figures, however this snapshot inherently overestimates the numbers of new cars and underestimates the number of old cars in relation to the average stock during the year. In order to provide a more balanced picture the stock figures incorporate factors aimed at giving a closer estimate of average stock. Further details of these adjustments can be found in Appendix 6.

2.3   The accident involvement and car user casualty rates in this section are given per 10,000 licensed vehicles in each group. General information on the drivers and average mileage covered by different types of car is available from the National Travel Survey and is summarised in Appendix 7. This shows that company cars record higher mileages than privately owned cars, and that newer cars record higher mileages than older cars.

2.4   The reliability of each accident involvement and casualty rate presented in Part 2 depends on the number of licensed cars in that category. For this reason, the rates are not quoted for car types with populations of less than 20,000 licensed vehicles.

Tables C : Description

2.5   Table C shows the extent to which drivers and their cars become involved in injury accidents. This is influenced mainly by driver behaviour, average mileage and pattern of use. The safety characteristics of the car will also have some influence but this is difficult to assess.

2.6   The table shows rates of involvement in injury accidents per 10,000 licensed vehicles in each of a number of groups. All injury accident involvements are included, regardless of whether the injured person was inside the car or not. Information is shown by size of car, private or company ownership, age of car, and by severity of injury.

2.7   The purpose is to illustrate differences in the rate of involvement in injury accidents for different groups of car. General information on car use, by age of car, engine capacity and ownership, extracted from the National Travel Survey, is summarised in Appendix 7.

Table C : Commentary

2.8   It is important to remember that differences in rates between groups in Table C will strongly reflect differences in the level and type of use and of driver behaviour, and should be interpreted carefully. The table shows the following:

-       Drivers of new privately owned cars are involved in less accidents than drivers of new company owned cars. This may largely be explained by the National Travel Survey data which shows that in general company owned cars record higher mileages than privately owned cars.

-       Drivers of privately owned high performance cars generally have higher rates of involvement in injury accidents than drivers of their standard performance counterparts.

-       This distinction is much less clear for company owned cars where standard and high performance cars have very similar involvement rates.

-       Newer private cars are seen to have lower rates of involvement in injury accidents than older cars, despite the evidence in Appendix 7 which shows that newer cars on average cover more miles each year.

Table D : Description

2.9   Table D shows car user casualty rates, which are influenced by the combined effects of accident involvement, accident type, the number of passengers and secondary safety features.

2.10   The table shows car user casualty rates per 10,000 licensed vehicles. Only persons injured while travelling inside the car are included in these rates. Information is shown by size of car, private or company ownership, and by severity of injury to occupant.

2.11   The average number of passengers carried has a much greater influence on the rates presented in Table D. However the accident reporting system does not require the police to record the number of uninjured passengers, so it is difficult to assess the size of this influence on the figures presented.

Table D : Commentary

2.12   Again, interpretation of the figures in this table depends on the relative mileages of the different types of car, their patterns of use and the behaviour of their drivers. The table shows the following:

-       In most groups about 9 to 12 car users are killed or seriously injured each year for every 10,000 licensed vehicles.

-       The car user casualty rates show a tendency to decline with increasing size of car.

-       Among older cars, company owned vehicles have generally lower casualty rates than privately owned vehicles.

-       Newer private cars have lower casualty rates than their older counterparts.

Table C   Rates of involvement in injury accidents: by size of car and ownership: 1993

	Privately owned		Company owned	
	Injury accident severity		Injury accident severity	
Age of car Performance Size	Fatal or serious	All	Fatal or serious	All
Regd. on or since 1.1.91				
Standard				
Small	21	136	31	240
Small/medium	19	131	35	243
Medium	20	119	27	175
Large	20	110	23	149
All standard	20	128	28	192
High				
Small	25	159	30	192
Small/medium	22	137	31	178
Medium	23	118	27	179
All high	23	136	29	180
Registered before 1.1.91				
Standard				
Small	22	146	24	155
Small/medium	25	158	24	175
Medium	25	153	24	158
Large	21	126	18	108
All standard	24	150	23	149
High				
Small	35	195	30	175
Small/medium	33	185	24	148
Medium	24	144	24	149
All high	30	173	24	152

Injury accident involvements per 10,000 licensed vehicles in each group

Table D Car user casualty rates: by size of car and ownership: 1993

	Privately owned		Company owned	
Age of car				
Performance	Injury severity		Injury severity	
Size	Fatal or serious	All	Fatal or serious	All
Regd. on or since 1.1.91				
Standard				
Small	11	97	15	162
Small/medium	9	84	14	136
Medium	10	72	10	85
Large	9	53	7	59
All standard	10	84	10	99
High				
Small	15	115	11	116
Small/medium	11	76	10	86
Medium	11	71	9	84
All high	12	86	9	87
Registered before 1.1.91				
Standard				
Small	13	109	12	104
Small/medium	13	106	10	97
Medium	11	89	9	77
Large	8	60	6	43
All standard	12	98	9	76
High				
Small	22	141	13	116
Small/medium	17	115	9	66
Medium	11	81	10	71
All high	16	109	10	75

Car user casualties per 10,000 licensed vehicles in each group

PART 3

CAR SAFETY RATINGS: RELATED ANALYSES

The aggressivity of cars in two-car collisions

3.1 In Part 1 it was noted that although mass was beneficial to a driver in a relatively heavy car, it could be detrimental to the drivers of other cars with which it collided. It is possible to calculate ratings based on the injuries sustained by the driver of the "other car" when colliding with certain make/models. In other words to calculate the proportion of drivers in Car 2 injured in collisions with a given model of car (Car 1).

3.2 Following the nomenclature in the description of Table A on page 4, the calculation of the risk of injury as the percentage of drivers injured in a car (say car 2) when colliding with a specific model of car (car 1) can be represented as follows:

$$\frac{X_2 + X_3}{X_1 + X_2 + X_3} \times 100$$

where

X_1 = Number of accidents in which driver injured in car 1 but not in car 2
X_2 = Number of accidents in which driver injured in car 1 and in car 2
X_3 = Number of accidents in which driver injured in car 2 but not in car 1

3.3 While, in general, models which were rated as safer than their group average were found to be more aggressive, this was not exclusively the case. There were examples of models offering, say, average protection but being significantly either more or less aggressive than average. In other words, although generally, as one would expect, the ratings for aggressivity were the inverse of those for safety, there were some significant differences both across the range of models and within size groups.

3.4 Chart E plots the best estimates of the aggressivity ratings against those for the risk of injury and illustrates that the former generally correlate highly with the inverse of the latter. The chart also contains one or two outliers suggesting that some models may not adhere to this pattern; for example, there were some models which perform relatively well on at least one of the measures without necessarily having an inverse (ie poor) rating on the other. In other words the aggregate number of injuries in collisions involving these models was less than average, suggesting that some models may strike a better balance than others between protecting their own drivers and minimising their aggressivity towards others.

Chart E The relationship between the Aggressivity and Risk of injury ratings

Aggressivity ratings

Risk of injury ratings

Mass-adjusted car safety ratings

3.5   From an individual driver's point of view the mass of car is certainly to be regarded as a safety factor, because the laws of physics determine that they will be safer in most accidents when they are driving a heavier or larger car. However it would be interesting to compare cars only on the basis of the secondary protection that is offered by their structure, design and secondary safety fittings. This requires a procedure to remove the influence of mass from the calculation of risk of driver injury. Initial work has been carried out to remove the specific effects of mass from the relative risks of driver injury so that relative safety ratings of car models can be made in terms of their inherent design features and secondary safety fittings. The methodology used is that described in the research work carried out by the Transport Research Laboratory and set out in their recent report (see Appendix 1). In short, the mass adjusted ratings derived from the risk of injury are based on the difference between the unadjusted rating and the "expected" rating for a car of that mass. In theory this difference is attributable to model specific design effects. The expected rating is obtained by fitting a linear model between the risk of injury and mass for each of the car models covered in the analysis. This produced the following relationship:

"Expected" risk (per 100 accs) = 103.3 - 0.044 Mass   (R^2 = 0.77),
implying that for every extra 100 KG, the risk of injury (per 100 accidents) falls by 4.4.

3.6   The expected risks are then subtracted from the unadjusted risks of injury as published in Table A (ie unadjusted for mass) to produce the Mass-adjusted Safety Index (MSI).

3.7   After mass adjustment, the pattern of larger models generally being safest no longer applies, although early indications suggest there are very few small cars among the models with the safest ratings. This may be due to the fact that a more extensive range of safety fittings are more likely to be found in larger cars than smaller cars.

3.8   An assumption lying behind all the recent car safety work, and one which is particularly important to this analysis, is that each make/model collides with a similar distribution of cars in terms of the size and, more importantly, mass of the second car involved. This was found to be true in terms of the four size groups based on length of car and we are now able to investigate whether the assumption is correct with regard to mass. Appendix 4 shows the distribution of collisions for each model and size group of car split by the mass of the collision partner and confirms that the assumption holds.

3.9   The calculation of mass adjusted safety ratings depends critically upon the quality and availability of detailed mass data for specific car models. The ratings may also be influenced by differences in passenger and luggage loading in cars. This analysis is still at a experimental stage and further progress is dependent on the extent to which the necessary mass data can be accumulated. Therefore it would not be appropriate to publish model specific ratings at this stage.

Collisions between cars and pedestrians

3.10   The following analysis is based on data from accidents involving injured pedestrians and one car in the years 1986-93. To set the analysis in context, this subset of accidents is responsible for a substantial number of casualties; in 1993 about 800 pedestrians were killed and 9000 seriously injured by cars. Casualties of all severities in 1993 in these types of accident were nearly 40,000.

3.11   Car safety analysis is based upon the proportion of drivers injured in two-car collisions. In this assessment of the aggressivity of cars and drivers towards pedestrians, all pedestrians are, by definition, injured. The approach therefore focuses on the proportions of the pedestrians that were seriously or fatally injured, and possible contributory factors.

3.12   However, a preliminary investigation of the aggressivity of younger and older car drivers can be deduced from a comparison of their pedestrian collisions in relation to their mileage driven. Table F shows the numbers and proportions of car drivers involved in collisions with pedestrians by age and sex of car driver.

3.13   The table shows that 73% of collisions between cars and pedestrians involve male drivers which broadly reflects the proportion of total car mileage driven by men. However, the data show that young drivers are involved in a disproportionately large number of collisions with pedestrians when the estimates of their mileage is taken into account. This effect is particularly strong for male drivers, where those aged between 16 and 24 account for 7 percent of mileage driven but 19 percent of collisions with pedestrians. This could be due to inexperience and lack of roadcraft. For example younger drivers, and male drivers, may drive too fast for the road conditions reducing their reaction time to dangerous traffic situations.

3.14   Table G shows the effect of various factors on the severity of injury to pedestrians hit by cars. The table shows the following:

Pedestrian age.

3.15   Older pedestrians are more likely to be killed or seriously injured. This effect is most marked in the over 55 age group, where pedestrians are about one and a third times more likely than average to be seriously injured, and three times more likely to be killed.

Point of impact on car.

3.16   Not surprisingly, impacts with the rear of a car are least severe, probably because in these collisions vehicles are moving slower than average. Frontal impacts on pedestrians are three times as likely to be fatal as side impacts.

Speed limit of road.

3.17   Although only a proxy for speed of impact, the table predictably shows that collisions on faster roads are more severe. At the extreme, a person hit on a 70mph road is shown to have a one in five chance of being killed, over ten times the likelihood of being killed on a 30mph road.

Driver age.

3.18   Driver age has no significant effect on the likelihood of a pedestrian being killed or seriously injured. The table does, however, show that pedestrians hit by cars driven by drivers in the younger age bands are most likely to be killed. This could be because younger drivers drive at inappropriate speeds in urban areas, the net effect of which is to relatively increase both the number of collisions with pedestrians (see Table F) and the relative severity of the collisions.

Driver sex.

3.19   Pedestrians hit by a car with a male driver are slightly more likely to be killed or seriously injured and are about one and a half times more likely to be killed than by a car with a female driver, which may again reflect on the driving behaviour of males compared to females.

Table F Car - Pedestrian collisions, mileage driven, by age and sex of driver[1]

		Number of collisions	Percent	Mileage driven (%)
Male	16-24	53303	19.2	6.7
	25-34	57161	20.6	18.1
	35-54	65713	23.6	34.0
	55+	27685	10.0	13.7
	All ages	203862	73.3	72.4
Female	16-24	18378	6.6	3.8
	25-34	23229	8.4	7.9
	35-54	26157	9.4	12.4
	55+	6355	2.3	3.4
	All ages	74119	26.7	27.6
All drivers		277981	100.0	100.0

1 Relates to accidents in the period 1986-93. Mileage data from National Travel Survey.

Table G Severity of injury to pedestrians hit by cars: by selected factors, 1986-93

Percentage of injured pedestrians that are seriously or fatally injured[1]

| Factor | Injury severity | |
	Fatal or serious	Fatal
Age of pedestrian		
0-15	25.0	0.8
16-24	24.8	1.1
25-34	25.6	1.3
35-54	28.1	2.3
55 or more	38.9	6.1
Age of driver		
16-24	32.3	2.9
25-34	28.6	2.2
35-54	26.6	1.8
55 or more	27.2	1.9
Sex of driver		
Male	29.2	2.4
Female	26.1	1.5
Point of impact with car		
Front	30.4	2.7
Back	17.1	0.5
Offside	25.0	0.9
Nearside	23.6	0.9
Speed limit of road (mph)		
20 or 30	26.4	1.6
40 or 50	42.0	6.1
60	44.1	7.7
70	59.7	19.1
All car-pedestrian collisions	27.9	2.1

1 Starting point is injured pedestrians.

APPENDIX 1   BACKGROUND AND DETAILED NOTES

Introduction

By agreement between the Department of Transport and police forces throughout Great Britain, details of all road accidents reported to the police in which a person is injured are transmitted to the Department of Transport in the form of a standard report (STATS19), usually through the appropriate local authority. These accident reports provide the basis for the Department's work on monitoring and analysis of road accident statistics.

From January 1989 the standard report prepared by the police was modified to include the registration marks of all motor vehicles involved in accidents. By linking the registration marks with the vehicle data held by DVLA at Swansea, a range of extra information about the vehicles involved in injury accidents can now be obtained. No personal details, such as names or addresses, are collected in this process.

This is the fifth report of results based on these additional vehicle data. It presents estimates of injury risk to drivers involved in injury accidents for popular car models, and also injury accident rates for different types of car and the types and numbers of casualties resulting from these accidents. The first report of this type, covering accidents in 1989, was published by HMSO in May 1991.

Coverage

It is not possible to link all vehicles involved in road injury accidents with vehicle registration data at DVLA. Details of foreign, diplomatic and military vehicles and those with trade plates are not held at DVLA. In addition, registration marks are generally unavailable for vehicles which leave the scene of an accident and some registration marks which appear to have a valid format prove untraceable.

Since 1991 these factors collectively resulted in roughly 15 per cent of vehicles having no additional vehicle details. Due to the varying rates at which registration marks began to be recorded by police forces the corresponding figures for preceding years were higher than this.

The number of two car collisions analysed to produce the tables in Part 1 of this report is considerably lower than the number of two car collisions recorded in, say, the Department's annual publication Road Accidents Great Britain. This difference is due to several factors. In order to be included in the analysis it is necessary to have the vehicle registration data of **both** cars involved. Although, for the reasons outlined in the preceding paragraphs, the detailed vehicle data is available for over 80 per cent of the cars, the requirement to have data for both cars results in approximately 70 per cent of the two car accidents (as published in Road Accidents Great Britain) being included in the analysis.

The number of accidents is further reduced due to several refinements to the dataset which aim to ensure that cars are being compared on as even a basis as possible. This involves narrowing the broader definition of car that is used for other publications and excluding accidents involving parked cars. In addition the analysis is restricted to accidents in which one or both of the drivers was injured. The resultant dataset, designed to minimise variation in accident types between different car models which could unduly bias comparisons of driver injury risk in particular car models, contains approximately 45 per cent of the two car accidents published in Road Accidents Great Britain. Accident types that could otherwise bias ratings are therefore directly eliminated from the analysis. If they had been retained the relatively less powerful procedure of having to allow for their influence in a more complex modelling process would have had to be adopted.

Any bias due to missing data would be unlikely to significantly influence the ratings. It is reasonable to expect that missing data are unlikely to be concentrated in particular types of accidents for particular models of car, but are more likely to be randomly spread over all accident types and all models of car. Tests were conducted to check that the final dataset is typical of all two car collisions in terms of driver and accident characteristics. The results, shown below, demonstrate that in terms of driver sex, driver age and speed limit of road the dataset used in the secondary safety analysis is broadly representative.

Two car collisions by accident/driver type (%), 1989-93

	Road Accidents GB Table 23	Cars: Make and Model Table A
Speed limit of road		
20-39	61	58
40-59	11	12
60-69	23	25
70	5	5
Total	100	100
Driver sex		
Male	69	63
Female	31	37
Total	100	100
Driver age		
17-24	29	29
25-34	28	28
35-54	31	31
55+	13	13
Total	100	100

Other publications

Similar reports on car safety have been published in the United States of America by the Insurance Institute for Highway Safety, in Sweden by the Folksam Insurance Company, in Australia by the Monash University Accident Research Centre, in Finland by the University of Oulu and in the UK by the Transport Research Laboratory. Each produces a summary report for widespread distribution.

References:-

Insurance Institute for Highway Safety: Status Report. Vol 24, No 11.
1005 North Glebe Road, Arlington, VA 22201. ISSN 0018-988X.

Safe and Dangerous Cars 1989-90: A report from Folksam.
Folksam, Division for Research and Development, S-106 60 Stockholm.

The Effect of Driver's Age and Experience and Car Model on Accident Risk.
University of Oulu, Finland, Publications of Road and Transport Laboratory 1992.

Vehicle Crashworthiness Ratings: Victoria 1983-90 and NSW 1989-90 Crashes. Technical Report.
Monash University Accident Research Centre 1992.

The Theoretical Basis for Comparing the Accident Record of Car Models (Project Report 70).
Dr J Broughton, Safety Research Centre, Transport Research Laboratory 1994.

APPENDIX 2   DEFINITIONS

The statistics refer to accidents involving cars resulting in personal injury on public roads, including footways, which became known to the police. Results for 1993 include all accidents in that year determined by the date of accident. Tables in Part 1 also include data from 1989, 1990, 1991 and 1992 accidents. Figures for deaths refer to persons who sustained injuries causing death at the time of the accident or within 30 days of the accident, which is the internationally recognised definition.

Injury severity	*Severity* of an *injury* to a casualty is determined by the degree of injury and is either *fatal, seriously injured* or *slightly injured.*
Accident Severity	*Severity* of an *accident*, is determined by the severity of injury of the most severely injured casualty in that accident. That is *fatal*, where one or more persons involved in the accident were killed, *serious*, where one or more persons involved in the accident were seriously injured but no-one killed, or *slight*, where one or more persons involved in the accident were slightly injured but no-one killed or seriously injured.
Car	Any four-wheeled car. This includes saloons, hatch-backs, estates, "people carriers", coupes and convertibles. Purpose-built taxis, car derived vans, goods vehicles and minibuses are not included.
Size	Not formally defined but arranged so that the consumer can recognise standard groups. As an approximate guide, cars in the *small* group are generally between 140 and 150 inches in length and broadly equate to the motor industry's "minis" and "super-minis". Those in the *small/medium* group (equal to the industry's "lower medium") are between 155 and 165 inches, those in the *medium* group (= "upper medium") between 170 and 180 inches, and those in the *large* group (= "executive" and "luxury") over 180 inches.
	The allocation of a particular model to a size group does not imply that the model meets any formal classification or standard.
High performance	No single criterion has been adopted for the purpose of identifying cars in this category. Models whose performance is considerably higher than the standard production range are included. Most will be fitted with engines of higher capacity than their standard performance counterparts, and may also have features such as fuel injection or be fitted with turbo chargers. Typically, though not invariably, they have the capability of accelerating from 0 to 60 mph in 10 seconds or less.
	In the large group most cars are fitted with engines of greater power and higher cubic capacity. There is less distinction between standard and higher performance cars, so all cars in this size group have been classed as standard performance.
New and *Old* cars	Used in this report to describe cars registered on or after 1st January 1991 and before 1st January 1991 respectively. That is *new*, to describe cars up to three years old at the end of 1993, and *old*, to

describe cars three years old or more at the same date.

Ownership type
Identified from the registered keeper record at DVLA. *Company* cars are registered in the name of a company or partnership. *Private* cars are registered in the name of an individual.

APPENDIX 3   NUMBER OF INVOLVEMENTS BY MAKE/MODEL IN TABLE A

Car size/model	Registration dates	Involvements	Car size/model	Registration dates	Involvements
SMALL			**SMALL/MEDIUM**		
CITROEN 2CV/DYANE	Jan 84 - Jul 90	338	ALFA ROMEO 33	Jan 84 - Dec 93	164
CITROEN AX	Jun 87 - Dec 93	1219	CITROEN ZX	Jun 91 - Dec 93	245
CITROEN VISA	Jan 84 - Jul 88	255	FIAT STRADA/REGATA	Jan 84 - Jun 88	548
FIAT 126	Jan 84 - Dec 92	152	FIAT TIPO/TEMPRA	Jul 88 - Dec 93	484
FIAT PANDA	Jan 84 - Dec 93	1351	FORD ESCORT/ORION	Jan 84 - Aug 90	22035
FIAT UNO	Jan 84 - Dec 93	2826	FORD ESCORT/ORION	Sep 90 - Dec 93	2487
FORD FIESTA	Jan 84 - Mar 89	10455	HONDA CIVIC	Oct 87 - Oct 91	354
FORD FIESTA	Apr 89 - Dec 93	4454	HYUNDAI PONY	Oct 85 - Aug 90	324
NISSAN MICRA	Jan 84 - Dec 92	3420	LADA RIVA[6]	Jan 84 - Dec 93	1095
PEUGEOT 106	Oct 91 - Dec 93	170	LADA SAMARA	Nov 87 - Dec 93	390
PEUGEOT 205	Jan 84 - Dec 93	4388	LANCIA DELTA/PRISMA	Jan 84 - Dec 93	171
RENAULT 5	Jan 84 - Jan 85	221	MAZDA 323	Jan 84 - Aug 85	160
RENAULT 5	Feb 85 - Dec 93	2131	MAZDA 323	Sep 85 - Sep 89	373
RENAULT CLIO	Mar 91 - Dec 93	312	MAZDA 323	Oct 89 - Dec 93	184
ROVER METRO	Jan 84 - Mar 90	8201	NISSAN CHERRY	Jan 84 - Aug 86	678
ROVER METRO	Apr 90 - Dec 93	1054	NISSAN SUNNY	Jan 84 - Aug 86	1021
ROVER MINI	Jan 84 - Dec 93	1683	NISSAN SUNNY	Sep 86 - Jan 91	1668
TALBOT SAMBA	Jan 84 - Sep 86	247	PEUGEOT 305	Jan 84 - Jul 88	427
VAUXHALL NOVA	Jan 84 - Mar 93	5297	PEUGEOT 309	Feb 86 - Mar 93	2398
VOLKSWAGEN POLO	Jan 84 - Dec 93	2269	PROTON 1.3/1.5	Mar 89 - Dec 93	330
YUGO 3/4/500	Jan 84 - Dec 91	175	RENAULT 19	Feb 89 - Dec 93	655
YUGO TEMPO	Jan 84 - Dec 93	385	RENAULT 9/11	Jan 84 - Jan 89	1068
			ROVER 200	Jun 84 - Sep 89	3008
TOTAL[1]		78310	ROVER 200/400	Oct 89 - Dec 93	1714
			ROVER MAESTRO	Jan 84 - Dec 93	3847
			SEAT IBIZA/MALAGA	Oct 85 - Sep 93	523
			SKODA ESTELLE	Jan 84 - Jul 90	667
			TALBOT HORIZON	Jan 84 - Dec 85	272
			TOYOTA COROLLA	Jan 84 - Aug 87	582
			TOYOTA COROLLA	Sep 87 - Jul 92	651
			VAUXHALL ASTRA/BELMONT	Oct 84 - Sep 91	8607
MEDIUM			VAUXHALL ASTRA	Oct 91 - Dec 93	713
AUDI 80/90	Jan 84 - Oct 86	628	VOLKSWAGEN GOLF/JETTA	Jan 84 - Feb 84	603
AUDI 80/90	Nov 86 - Dec 93	468	VOLKSWAGEN GOLF/JETTA	Mar 84 - Jan 92	3979
BMW 3 SERIES	Jan 84 - Mar 91	1985	VOLKSWAGEN GOLF/VENTO	Feb 92 - Dec 93	215
BMW 3 SERIES	Apr 91 - Dec 93	215	VOLVO 300	Jan 84 - Dec 91	2089
CITROEN BX	Jan 84 - Dec 93	2545			
FORD SIERRA/SAPPHIRE	Jan 84 - Dec 93	13659	TOTAL[1]		88633
FSO POLONEZ	Jan 84 - Dec 91	181			
HONDA ACCORD	Oct 85 - Sep 91	587			
HONDA PRELUDE	Jan 84 - Mar 92	185	**LARGE**		
HYUNDAI STELLAR	Jun 84 - Dec 91	278	AUDI 100/200	Jan 84 - Dec 93	402
MAZDA 626	Jan 84 - Sep 87	312	BMW 5 SERIES	Jan 84 - May 88	443
MAZDA 626	Oct 87 - Jan 92	286	BMW 5 SERIES	Jun 88 - Dec 93	458
MERCEDES 190	Jan 84 - Sep 93	605	FORD GRANADA	Jan 84 - Apr 85	407
NISSAN BLUEBIRD	Mar 86 - Aug 90	2493	FORD GRANADA	May 85 - Dec 93	2178
NISSAN PRIMERA	Sep 90 - Dec 93	212	JAGUAR XJ	Oct 86 - Dec 93	511
NISSAN STANZA	Jan 84 - Dec 86	235	MERCEDES S CLASS	Jan 84 - Sep 91	174
PEUGEOT 405	Jan 88 - Dec 93	1992	MERCEDES 200/300	Jan 84 - Sep 85	161
RENAULT 18	Jan 84 - May 86	180	MERCEDES 200/300	Oct 85 - Dec 93	522
RENAULT 21	Jun 86 - Dec 93	969	PEUGEOT 505	Jan 84 - Dec 91	265
ROVER MONTEGO	Apr 84 - Dec 93	4625	RENAULT 25	Jul 84 - Jan 93	592
SUBARU 1.6/1.8	Nov 84 - Dec 91	193	ROVER 800	Jul 86 - Dec 93	1453
TALBOT ALPINE/SOLARA	Jan 84 - Dec 86	254	ROVER SD1	Jan 84 - Jun 86	320
TOYOTA CAMRY	Jan 84 - Dec 86	246	SAAB 900	Jan 84 - Sep 93	428
TOYOTA CARINA	Apr 84 - Feb 88	342	SAAB 9000	Oct 85 - Dec 93	263
TOYOTA CARINA	Mar 88 - Apr 92	325	VAUXHALL CARLTON	Jan 84 - Oct 86	595
VAUXHALL CAVALIER	Jan 84 - Sep 88	5655	VAUXHALL CARLTON	Nov 86 - Dec 93	1191
VAUXHALL CAVALIER	Oct 88 - Dec 93	4972	VAUXHALL SENATOR	Sep 87 - Dec 93	321
VOLKSWAGEN PASSAT/SANTANA	Jan 84 - May 88	447	VOLVO 200	Jan 84 - Dec 93	667
VOLKSWAGEN PASSAT	Jun 88 - Dec 93	363	VOLVO 700	Jan 84 - Jul 91	1157
VOLVO 400	Jun 87 - Dec 93	532			
TOTAL[1]		83261	TOTAL[1]		19879

1 Totals include listed models, models whose sample size was insufficient and models whose first registration date is earlier than 1 January 1984.

APPENDIX 4 DISTRIBUTION OF INVOLVEMENTS BY MASS OF COLLISION PARTNER

Car size/model	Registration dates	Collision Partner Mass Group (%)										
		<600 Kg	601-700 Kg	701-800 Kg	801-900 Kg	901-1000 Kg	1001-1100 Kg	1101-1200 Kg	1201-1300 Kg	1301-1400 Kg	>1400 Kg	Total
SMALL												
CITROEN 2CV/DYANE	Jan 84 - Jul 90	2	0	9	24	33	14	11	1	6	1	100
CITROEN AX	Jun 87 - Dec 93	0	0	8	24	33	16	11	2	4	2	100
CITROEN VISA	Jan 84 - Jul 88	0	0	9	21	35	14	12	2	5	3	100
FIAT 126	Jan 84 - Dec 92	0	1	8	24	33	13	12	1	8	1	100
FIAT PANDA	Jan 84 - Dec 93	0	0	9	25	31	13	12	3	4	2	100
FIAT UNO	Jan 84 - Dec 93	0	0	8	24	33	14	12	2	5	2	100
FORD FIESTA	Jan 84 - Mar 89	0	0	8	26	32	14	12	2	5	2	100
FORD FIESTA	Apr 89 - Dec 93	0	0	7	26	32	14	12	2	5	2	100
NISSAN MICRA	Jan 84 - Dec 92	0	0	9	26	29	15	11	2	5	2	100
PEUGEOT 106	Oct 91 - Dec 93	0	0	6	22	34	20	10	2	5	1	100
PEUGEOT 205	Jan 84 - Dec 93	0	0	9	26	30	14	11	2	6	2	100
RENAULT 5	Jan 84 - Jan 85	0	0	8	26	29	14	11	3	6	3	100
RENAULT 5	Feb 85 - Dec 93	0	0	8	26	30	15	11	2	6	2	100
RENAULT CLIO	Mar 91 - Dec 93	0	0	12	22	32	15	10	3	4	2	100
ROVER METRO	Jan 84 - Mar 90	0	0	8	26	31	14	11	2	5	2	100
ROVER METRO	Apr 90 - Dec 93	0	0	8	25	29	16	12	2	6	2	100
ROVER MINI	Jan 84 - Dec 93	0	0	8	25	30	17	11	2	5	2	100
TALBOT SAMBA	Jan 84 - Sep 86	1	0	9	27	29	16	11	2	4	1	100
VAUXHALL NOVA	Jan 84 - Mar 93	0	0	9	24	31	16	11	2	5	2	100
VOLKSWAGEN POLO	Jan 84 - Dec 93	0	0	9	25	29	15	11	2	5	2	100
YUGO 3/4/500	Jan 84 - Dec 91	0	0	7	32	28	17	8	1	7	1	100
YUGO TEMPO	Jan 84 - Dec 93	0	0	7	29	34	13	10	1	3	1	100
ALL SMALL		0	0	8	25	31	15	11	2	5	2	100
SMALL/MEDIUM												
ALFA ROMEO 33	Jan 84 - Dec 93	2	0	6	23	38	15	11	1	3	2	100
CITROEN ZX	Jun 91 - Dec 93	0	0	10	22	28	14	16	2	4	2	100
FIAT STRADA/REGATA	Jan 84 - Jun 88	0	0	9	27	32	14	10	2	3	2	100
FIAT TIPO/TEMPRA	Jul 88 - Dec 93	0	0	8	25	32	14	11	1	5	2	100
FORD ESCORT/ORION	Jan 84 - Aug 90	0	0	8	26	32	13	11	2	5	2	100
FORD ESCORT/ORION	Sep 90 - Dec 93	0	0	8	27	33	14	11	2	4	1	100
HONDA CIVIC	Oct 87 - Oct 91	0	0	10	24	29	15	13	3	5	1	100
HYUNDAI PONY	Oct 85 - Aug 90	0	0	11	26	28	17	9	2	6	0	100
LADA RIVA[6]	Jan 84 - Dec 93	0	0	9	25	31	15	11	2	5	2	100
LADA SAMARA	Nov 87 - Dec 93	0	0	9	30	30	13	10	2	4	1	100
LANCIA DELTA/PRISMA	Jan 84 - Dec 93	1	1	6	25	34	16	12	2	4	1	100
MAZDA 323	Jan 84 - Aug 85	0	0	8	25	29	13	15	2	5	2	100
MAZDA 323	Sep 85 - Sep 89	1	1	9	27	26	13	15	2	4	2	100
MAZDA 323	Oct 89 - Dec 93	0	0	14	26	27	9	15	1	6	2	100
NISSAN CHERRY	Jan 84 - Aug 86	0	0	9	26	30	15	11	3	4	2	100
NISSAN SUNNY	Jan 84 - Aug 86	0	0	8	26	32	13	12	2	5	2	100
NISSAN SUNNY	Sep 86 - Jan 91	0	0	9	25	31	13	12	3	4	2	100
PEUGEOT 305	Jan 84 - Jul 88	1	0	9	27	30	12	10	2	7	1	100
PEUGEOT 309	Feb 86 - Mar 93	0	0	9	26	31	15	11	2	5	2	100
PROTON 1.3/1.5	Mar 89 - Dec 93	0	0	9	26	31	15	8	1	7	1	100
RENAULT 19	Feb 89 - Dec 93	0	0	8	26	32	15	11	1	5	1	100
RENAULT 9/11	Jan 84 - Jan 89	0	0	9	25	31	15	11	2	5	2	100
ROVER 200	Jun 84 - Sep 89	0	0	10	27	30	14	11	2	5	1	100
ROVER 200/400	Oct 89 - Dec 93	0	0	10	26	30	15	10	2	5	2	100
ROVER MAESTRO	Jan 84 - Dec 93	0	0	8	28	31	14	11	2	5	2	100
SEAT IBIZA/MALAGA	Oct 85 - Sep 93	0	0	9	27	32	15	11	1	5	1	100
SKODA ESTELLE	Jan 84 - Jul 90	0	0	8	28	32	13	12	2	4	1	100
TALBOT HORIZON	Jan 84 - Dec 85	0	0	9	24	29	15	11	3	5	3	100
TOYOTA COROLLA	Jan 84 - Aug 87	0	0	9	27	29	14	12	3	4	2	100
TOYOTA COROLLA	Sep 87 - Jul 92	0	0	7	25	31	15	14	2	3	2	100
VAUXHALL ASTRA/BELMONT	Oct 84 - Sep 91	0	0	9	26	30	16	10	2	5	2	100
VAUXHALL ASTRA	Oct 91 - Dec 93	0	0	10	27	28	16	11	2	4	1	100
VOLKSWAGEN GOLF/JETTA	Jan 84 - Feb 84	0	0	9	26	29	15	11	2	5	2	100
VOLKSWAGEN GOLF/JETTA	Mar 84 - Jan 92	0	0	9	27	29	15	10	2	6	2	100
VOLKSWAGEN GOLF/JETTA	Feb 92 - Dec 93	0	0	7	28	32	15	10	1	4	3	100
VOLVO 300	Jan 84 - Dec 91	0	0	10	26	32	15	9	2	4	1	100
ALL SMALL/MEDIUM		0	0	9	26	31	14	11	2	5	2	100

APPENDIX 4 DISTRIBUTION OF INVOLVEMENTS BY MASS OF COLLISION PARTNER (CONT.)

Car size/model	Registration dates	<600 Kg	601-700 Kg	701-800 Kg	801-900 Kg	901-1000 Kg	1001-1100 Kg	1101-1200 Kg	1201-1300 Kg	1301-1400 Kg	>1400 Kg	Total
MEDIUM												
AUDI 80/90	Jan 84 - Oct 86	0	0	10	26	30	13	12	2	5	2	100
AUDI 80/90	Nov 86 - Dec 93	0	0	10	28	29	12	13	2	5	2	100
BMW 3 SERIES	Jan 84 - Mar 91	0	0	9	26	30	13	12	2	6	2	100
BMW 3 SERIES	Apr 91 - Dec 93	1	0	12	25	26	11	15	1	7	3	100
CITROEN BX	Jan 84 - Dec 93	0	0	9	26	33	12	12	2	5	1	100
FORD SIERRA/SAPPHIRE	Jan 84 - Dec 93	0	0	9	26	31	13	12	2	5	2	100
FSO POLONEZ	Jan 84 - Dec 91	0	1	10	27	29	16	8	3	6	1	100
HONDA ACCORD	Oct 85 - Sep 91	0	0	8	29	28	15	11	2	5	3	100
HONDA PRELUDE	Jan 84 - Mar 92	1	0	6	30	26	16	16	2	2	2	100
HYUNDAI STELLAR	Jun 84 - Dec 91	0	0	8	29	29	16	14	1	3	1	100
MAZDA 626	Jan 84 - Sep 87	0	0	10	29	30	14	8	3	5	1	100
MAZDA 626	Oct 87 - Jan 92	0	0	6	27	32	13	13	1	6	2	100
MERCEDES 190	Jan 84 - Sep 93	0	0	9	29	29	12	12	2	5	3	100
NISSAN BLUEBIRD	Mar 86 - Aug 90	0	0	9	28	30	13	12	3	4	1	100
NISSAN PRIMERA	Sep 90 - Dec 93	1	0	8	28	25	18	12	2	5	2	100
NISSAN STANZA	Jan 84 - Dec 86	0	0	8	31	28	12	9	3	6	2	100
PEUGEOT 405	Jan 88 - Dec 93	0	0	9	27	31	14	11	2	5	1	100
RENAULT 18	Jan 84 - May 86	0	0	8	29	29	14	10	3	6	1	100
RENAULT 21	Jun 86 - Dec 93	0	0	11	27	29	13	11	3	5	1	100
ROVER MONTEGO	Apr 84 - Dec 93	0	0	10	26	30	15	10	2	5	2	100
SUBARU 1.6/1.8	Nov 84 - Dec 91	1	0	11	28	29	11	10	2	6	2	100
TALBOT ALPINE/SOLARA	Jan 84 - Dec 86	0	0	8	28	29	16	10	2	5	2	100
TOYOTA CAMRY	Jan 84 - Dec 86	0	0	10	25	30	18	11	3	3	2	100
TOYOTA CARINA	Apr 84 - Feb 88	0	0	11	27	28	13	12	1	4	3	100
TOYOTA CARINA	Mar 88 - Apr 92	0	0	8	34	29	11	11	1	5	0	100
VAUXHALL CAVALIER	Jan 84 - Sep 88	0	0	9	26	31	15	10	1	5	1	100
VAUXHALL CAVALIER	Oct 88 - Dec 93	0	0	9	27	28	17	11	1	5	2	100
VOLKSWAGEN PASSAT/SANTANA	Jan 84 - May 88	0	0	8	27	30	15	10	2	6	1	100
VOLKSWAGEN PASSAT	Jun 88 - Dec 93	0	0	9	30	24	15	13	2	4	2	100
VOLVO 400	Jun 87 - Dec 93	0	0	10	26	33	17	9	1	3	1	100
ALL MEDIUM		0	0	9	27	30	14	11	2	5	2	100
LARGE												
AUDI 100/200	Jan 84 - Dec 93	0	0	11	27	32	10	11	2	4	2	100
BMW 5 SERIES	Jan 84 - May 88	0	0	11	24	32	14	11	3	4	2	100
BMW 5 SERIES	Jun 88 - Dec 93	1	0	9	30	28	14	11	2	6	1	100
FORD GRANADA	Jan 84 - Apr 85	0	0	9	30	31	12	11	2	3	2	100
FORD GRANADA	May 85 - Dec 93	0	0	9	26	31	13	13	1	5	2	100
JAGUAR XJ	Oct 86 - Dec 93	0	0	10	26	30	12	14	3	4	2	100
MERCEDES S CLASS	Jan 84 - Sep 91	0	0	7	27	32	12	14	0	5	3	100
MERCEDES 200/300	Jan 84 - Sep 85	0	0	10	28	28	12	12	1	6	1	100
MERCEDES 200/300	Oct 85 - Dec 93	0	0	12	26	31	13	12	1	4	1	100
PEUGEOT 505	Jan 84 - Dec 91	0	0	10	23	30	12	14	2	7	2	100
RENAULT 25	Jul 84 - Jan 93	0	0	8	29	29	15	11	2	4	2	100
ROVER 800	Jul 86 - Dec 93	0	0	10	29	28	16	10	2	3	1	100
ROVER SD1	Jan 84 - Jun 86	0	0	10	27	32	15	9	1	4	1	100
SAAB 900	Jan 84 - Sep 93	0	0	10	26	26	18	11	1	5	2	100
SAAB 9000	Oct 85 - Dec 93	1	0	5	28	34	13	12	3	2	2	100
VAUXHALL CARLTON	Jan 84 - Oct 86	0	0	9	26	32	15	10	2	4	1	100
VAUXHALL CARLTON	Nov 86 - Dec 93	0	0	9	27	31	13	12	1	4	1	100
VAUXHALL SENATOR	Sep 87 - Dec 93	0	0	11	28	30	14	10	3	2	2	100
VOLVO 200	Jan 84 - Dec 93	0	0	9	28	31	14	9	2	5	1	100
VOLVO 700	Jan 84 - Jul 91	0	0	9	29	28	13	13	1	5	1	100
ALL LARGE		0	0	9	28	30	14	11	2	4	2	100

APPENDIX 5   EXPLANATORY NOTES TO STATISTICAL MODELLING IN TABLES A AND B

The risk of injury to a driver involved in an accident clearly depends on a range of factors. Table B shows that the age and sex of driver, the speed of road and point of impact can all effect the risk of injury.

Valid comparisons of the inherent secondary safety records of different models of car are therefore complicated. Involvement in a high proportion of accidents with a low risk of driver injury will tend to produce a better apparent safety record whereas involvement in a high proportion of accidents with a high risk of driver injury will tend to produce a worse apparent safety record.

In order to isolate individual influences, statistical modelling has been used, which works by imagining that the risk of injury depends on a range of factors. The models look at the influences on the logarithm of the odds of injury, sometimes called the logit function, rather than directly at the influences on the proportion of accidents resulting in injury. This technique, known as logistic regression or logistic analysis, allows the influences on the risk of injury to be isolated and treated in an additive way. Unfortunately, the resulting factors are difficult to interpret by a non-specialist readership, so they have therefore been used to calculate corrected estimates of the percentage of drivers injured.

The statistical analyses used for Table B form the basis for the make/model comparisons in Table A. The largest influences on risk of injury to the driver in an accident identified from the statistical models used for Table B are also included in the statistical analyses comparing models of car used for Table A. The results are the best available estimates of the underlying secondary safety record, corrected for differences in key accident circumstances that influence the risk of injury to the driver.

The reliability of the result for a model of car in Table A depends on the number of recorded accident involvements of that model. Only those models exceeding the threshold of 150 involvements are included. In order to indicate the reliability of figures relating to individual models all estimates are accompanied by their associated 95 per cent confidence interval.

APPENDIX 5   EXPLANATORY NOTES TO STATISTICAL MODELLING (CONT.)

Statistical model used for each injury severity risk in Table A:

$$Y_{jklmn} = \mu + B_j + C_k + D_l + E_m + M_n + \varepsilon_{jklmn},$$

where $Y_{jklmn} = \text{Log}_e [P_{jklmn}/(1-P_{jklmn})]$

and $P_{jklmn} = n_{jklmn}/N_{jklmn},$

where $N_{jklmn} = $ Number of drivers in group $_{jklmn}$

and $n_{jklmn} = $ Number of injured drivers in group $_{jklmn},$

and where the effects are represented by: -

μ	Overall mean,
B_j	Speed limit of road,
C_k	First point of impact,
D_l	Sex of driver,
E_m	Age group of driver,
M_n	Effect for model of car, and
ε	Error term.

Statistical model used for each injury severity risk in Table B:

$$Y_{ijklm} = \mu + A_i + B_j + C_k + D_l + E_m + \varepsilon_{ijklm},$$

where the effects are represented as above and by: -

A_i	Size of car.

APPENDIX 6   ADJUSTMENTS TO 1993 STOCK FIGURES USED IN PART 2

The rates in tables C and D reflect an improvement to the calculation of stock levels. It is felt that the stock figures which act as the denominator in the calculations for involvement and casualty rates should be an average rather than an end of year figure. In the case of new cars (ie those registered from January 1991) using the end of 1993 stock figure would include all new registrations during that year, clearly constituting an overestimation of new cars relative to the average for the year.

There is a similar effect in the opposite direction for old cars (ie those registered before January 1991) due to scrappage. The transfer of cars between private and company ownership is an additional influence in the calculation of these factors, for example the transfer of older company cars to private ownership partly offsets the effect of scrappage on the figure for old private cars but accentuates the effect on the corresponding figure for company cars. The factors have been calculated using information from the Department's Vehicle Information Database.

Estimated Average Stock (1993) as a proportion of end of year stock:

Registered on or since 1.1.91

Privately owned: 0.811

Company owned: 0.857

Registered before 1.1.91

Privately owned: 1.035

Company owned: 1.329

APPENDIX 7   INFORMATION ON AVERAGE MILEAGE DRAWN FROM THE NATIONAL TRAVEL SURVEY 1991-93

1. Estimated annual average mileage of four wheeled cars, by age and ownership of car, 1991-93

	Company	Private
Up to 1 year old	22600	10400
Over 1 and up to 2 years old	20900	10100
Over 2 and up to 3 years old	20000	9900
Over 3 years old	15400	8100
All ages	19700	8500

2. Estimated annual average mileage of four wheeled cars, by engine capacity of car, 1991-93

Up to 1000 cc	6600
1001 to 1300 cc	7600
1301 to 1400 cc	9500
1401 to 1800 cc	10600
1801 to 2000 cc	13000
2001 cc or more	11100
All capacities	9500

Printed in the United Kingdom for HMSO

Dd 300473   C7   2/95   312257